Be An Expert!™

Mighty Trains

Erin Kelly

Children's Press®
An imprint of Scholastic Inc.

Contents

Know the Names

Be an expert! Get to know the names of these trains.

Steam Trains

Here come the trains!
Chug, chug, chug.

Zoom In

Find these parts on the steam train.

chimney

steam whistle

cab

wheel

steam

Freight Trains

They pull big loads.

locomotive

Expert Fact

The **locomotive** makes the train go. It pulls the other cars!

Passenger Trains

They carry people.

Let's take a trip.

Train Your Brain

Q: What happens at night on a **passenger** train?

A: Beds fold down so you can sleep.

High-Speed Trains

They go fast. Whoosh!

Acela

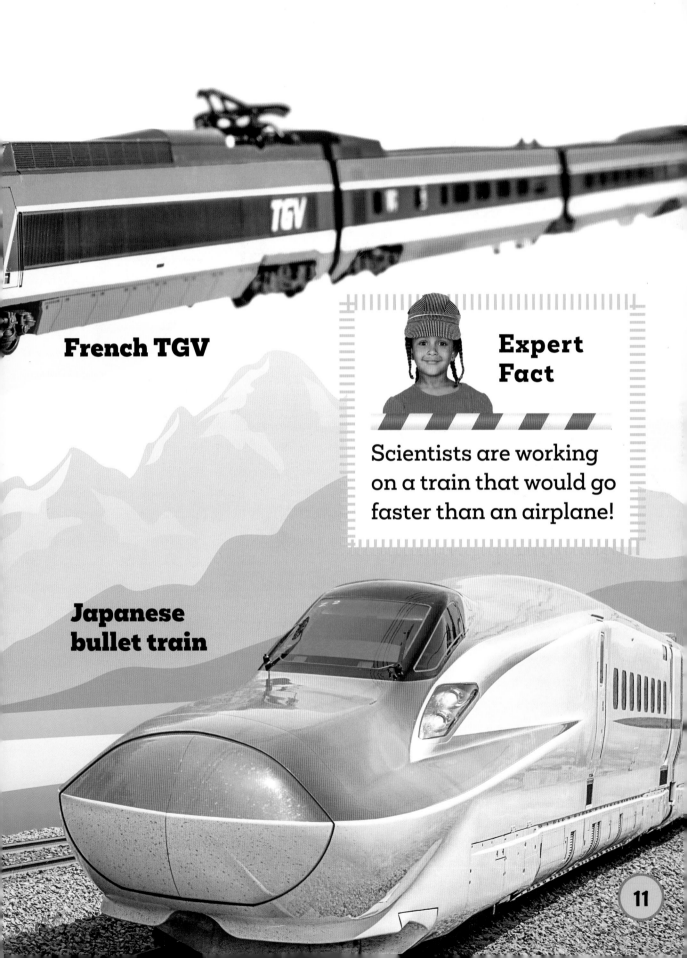

French TGV

Japanese bullet train

Expert Fact

Scientists are working on a train that would go faster than an airplane!

Subway Trains

They go underground.

Zoom In

Find these parts in the big picture.

train number | **lights** | **windshield wiper**

Monorails

They ride on one rail.

Expert Fact

Monorails are used in airports. You can find them at Disney World too!

hanging monorail

monorail

Mountain Trains

They go up, up, up!

Train Your Brain

Q: How does a mountain train climb steep hills?

A: The train has wheels with teeth. They hold on to the track.

Snow Trains

They move snow.

The **tracks** are clear now!

Expert Fact

A snow train can move a pile of snow that's as tall as a giraffe!

All the Trains

That was a good trip. Thanks, trains!

1.

2.

5.

6.

Expert Quiz

Do you know the names of these trains? Then you are an expert! See if someone else can name them too!

3.

4.

7.

8.

Answers: 1. Steam train. 2. Monorail. 3. Passenger train. 4. Freight train. 5. Subway train. 6. Snow train. 7. Mountain train. 8. High-speed train.

21

Expert Gear

Meet a freight train **conductor**. What does she have to do her job?

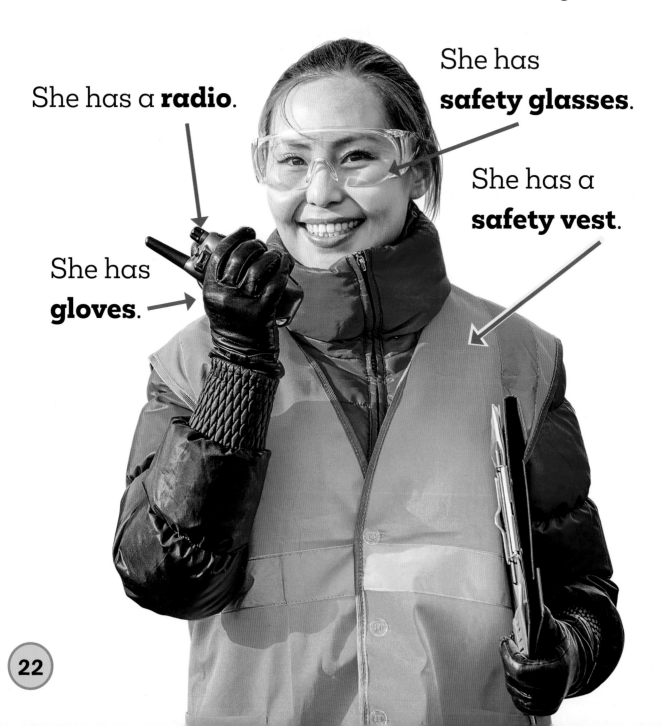

She has a **radio**.

She has **safety glasses**.

She has a **safety vest**.

She has **gloves**.

Glossary

conductor (kuhn-DUHK-tur):
the person in charge
of the train and the people.

locomotive (loh-kuh-MOH-tiv):
the engine that pulls the train.

passenger (pas-UHN-jer):
a traveler riding on a train.

tracks (TRAKS): the rails that
trains ride on.

Index

Library of Congress Cataloging-in-Publication Data

Names: Kelly, Erin Suzanne, 1965- author.

Title: Mighty trains / Erin Kelly.

Description: New York : Children's Press, an imprint of Scholastic Inc. 2020. | Series: Be an expert! | Includes index. | Audience: Grades K-1 | Summary: "Book introduces the reader to mighty trains"-- Provided by publisher.

Identifiers: LCCN 2019028554 | ISBN 9780531127612 (library binding) | ISBN 9780531132418 (paperback)

Subjects: LCSH: Railroad trains--Juvenile literature. | Monorail railroads--Juvenile literature.

Classification: LCC TF148 .K37 2020 | DDC 625.2--dc23

Printed in Heshan, China 62

SCHOLASTIC, CHILDREN'S PRESS, BE AN EXPERT™, and associated logos are trademarks and/or registered trademarks of Scholastic Inc.

2 3 4 5 6 7 8 9 10 R 29 28 27 26 25 24 23 22 21 20

Scholastic Inc., 557 Broadway, New York, NY 10012.

Art direction and design by THREE DOGS DESIGN LLC.

Photos ©: cover train: manwolste/iStockphoto; cover girl and throughout: CreativaImages/iStockphoto; cover hat and throughout: Adogslifephoto/Dreamstime; back cover train and throughout: Kenneth Sponsler/Dreamstime; spine and throughout: TheArtist/iStockphoto; 1 main: den-belitsky/iStockphoto; 1 sign: lpweber/iStockphoto; 2 top left and throughout: Sjo/iStockphoto; 2 center right and throughout: Vanessa Carvalho/Shutterstock; 2 bottom left girl and throughout: Ljupco/iStockphoto; 2 bottom left hat and throughout: Yoon S. Byun/The Boston Globe/Getty Images; 2 bottom right: Matthias Scholz/Alamy Images; 3 top left and throughout: Stephen B. Goodwin/Shutterstock; 3 top right and throughout: Sjo/iStockphoto; 3 center left and throughout: Christian Ohde/imageBROKER/agefotostock; 3 center right and throughout: Redwood8/Dreamstime; 3 bottom right and throughout: Besiki Kavtaradze/iStockphoto; 4 boy and throughout: kiankhoon/iStockphoto; 4 hat and throughout: Hemera Technologies/Getty Images; 4 flowers: Evgeniya Chertova/iStockphoto; 4-5 sky: almoond/iStockphoto; 6-7 background and throughout: MerggyR/iStockphoto; 6-7 grass: Anan Kaewkhammul/Shutterstock; 7 girl and throughout: Gelpi/Shutterstock; 7 hat and throughout: PhotoEuphoria/iStockphoto; 8 background and throughout: Nataniil/iStockphoto; 9 boy and throughout: sonyae/iStockphoto; 9 conductor and throughout: Julie Quarry/Alamy Images; 10 acela: Albert Pego/Shutterstock; 11 bullet train: Koichi Kumagai/AFLO/agefotostock; 12 wall: Mai Vu/iStockphoto; 14-15 background: Artis777/iStockphoto; 15 top right: José Fuste Raga/agefotostock; 17 man: Monty Rakusen/Getty Images; 17 wheel: Tamas Szabo/Wikimedia; 18-19 background: Yayasya/iStockphoto; 20 bottom left and throughout: Torresigner/iStockphoto; 22: Mark Agnor/Shutterstock; 23 girl on train: FamVeld/Shutterstock.